MW01359416

LIGHT

SUPER COOL SCIENCE EXPERIMENTS: LIGHT

by Dana Meachen Rau

CHERRY LAKE PUBLISHING • ANN ARBOR, MICHIGAN

A NOTE TO PARENTS AND TEACHERS: Please review the instructions for these experiments before your children do them. Be sure to help them with any experiments you do not think they can safely conduct on their own.

A NOTE TO KIDS: Be sure to ask an adult for help with these experiments when you need it. Always put your safety first!

Published in the United States of America by
Cherry Lake Publishing
Ann Arbor, Michigan
www.cherrylakepublishing.com

Content Editor: Robert Wolffe, EdD,
Professor of Teacher Education,
Bradley University, Peoria, Illinois

Book design and illustration: The Design Lab

Photo Credits: Cover and page 1, ©Alphaspirit/Dreamstime.com; page 8, ©iStockphoto.com/ldambies; page 12, ©Jeff R. Clow, used under license from Shutterstock, Inc.; page 16, ©aceshot1, used under license from Shutterstock, Inc.; page 17, ©Dgool/Dreamstime.com; page 21, ©Creatista/Dreamstime.com; page 24, ©iStockphoto.com/Argument; page 28, ©iStockphoto.com/bamby-bhamby

Library of Congress Cataloging-in-Publication Data
Rau, Dana Meachen, 1971–
Super cool science experiments: light / by Dana Meachen Rau.
p. cm.—(Science explorer)
Includes bibliographical references and index.
ISBN-13: 978-1-60279-531-0 ISBN-10: 1-60279-531-2 (lib. bdg.)
ISBN-13: 978-1-60279-610-2 ISBN-10: 1-60279-610-6 (pbk.)
1. Optics—Experiments—Juvenile literature. 2.Light—Experiments—Juvenile literature. I. Title. II. Series.
QC360.R38 2010
535.078—dc22 2009002701

Cherry Lake Publishing would like to acknowledge the work of The Partnership for 21st Century Skills. Please visit *www.21stcenturyskills.org* for more information.

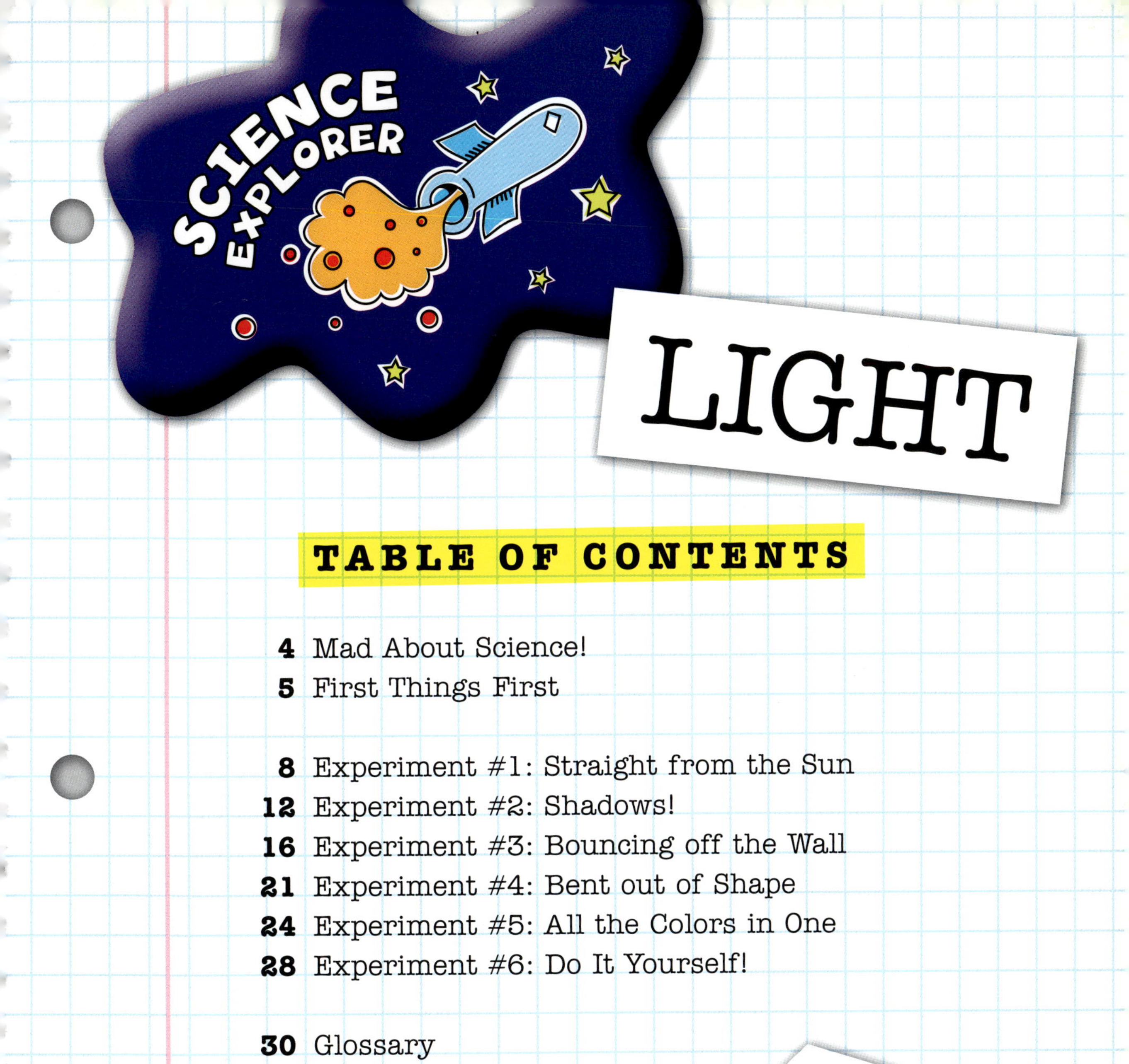

TABLE OF CONTENTS

4 Mad About Science!
5 First Things First

8 Experiment #1: Straight from the Sun
12 Experiment #2: Shadows!
16 Experiment #3: Bouncing off the Wall
21 Experiment #4: Bent out of Shape
24 Experiment #5: All the Colors in One
28 Experiment #6: Do It Yourself!

30 Glossary
31 For More Information
32 Index
32 About the Author

Notes:

Mad About Science!

Let's be scientists and discover things about light.

When you think of science, what comes to your mind? That only grown-ups do experiments? That experiments must be hard to do? You probably know that science can be interesting and fun. Would you believe that you can do experiments with things you already have at home? This book will help us learn how scientists think. We'll do that by experimenting with light. We don't have to be mad scientists to design our own experiments!

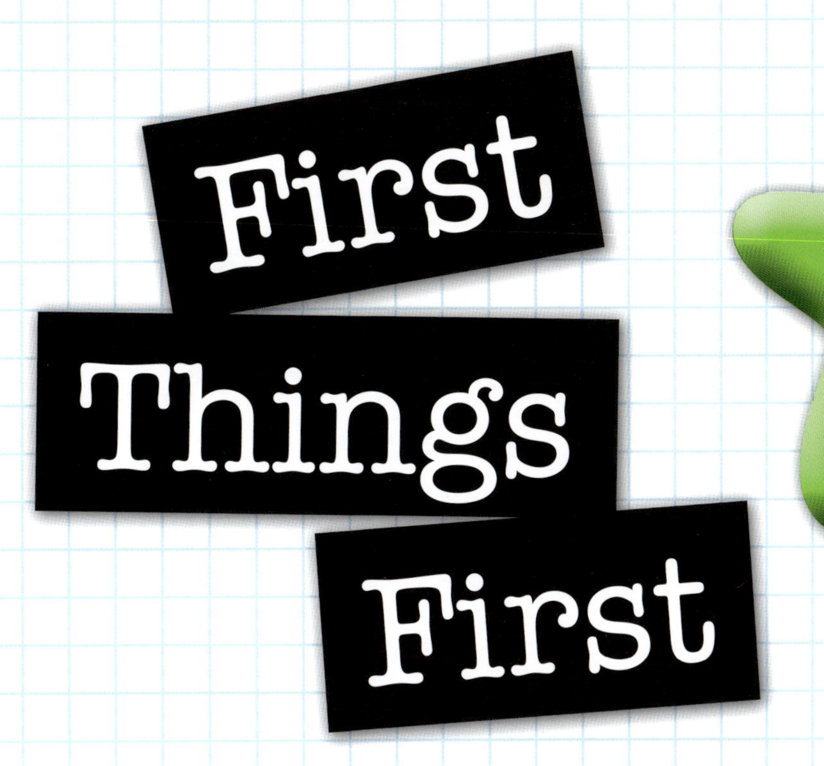

First Things First

How does light reflect off a shiny coin?

Scientists learn by studying things in nature very carefully. For example, scientists who study light compare sources of light. They look at the way light travels. They watch how light reflects off surfaces. Scientists do experiments to change the path of light. They test ways to break down light into parts, or to see if different light sources give off the same type of light.

Scientists take notes about what they observe.

Good scientists take notes on everything they discover. They write down their observations. Sometimes those observations lead scientists to ask new questions. With new questions in mind, they design new experiments to find the answers.

When scientists design experiments, they must think very clearly. The way they think about problems is often called the scientific method. What is the scientific method? It's a step-by-step way of finding answers to specific questions. The steps don't always follow the same pattern. Sometimes scientists change their minds. The process often works something like this:

Scientific Method

- **Step One:** A scientist gathers all the facts and makes observations about one particular thing.
- **Step Two:** The scientist comes up with a question that is not answered by all the observations and facts.
- **Step Three:** The scientist creates a hypothesis. This is a statement of what the scientist thinks is probably the answer to the question.

- **Step Four:** The scientist tests the hypothesis. He or she designs an experiment to see whether the hypothesis is correct. The scientist does the experiment and writes down what happens.
- **Step Five:** The scientist draws a conclusion based on how the experiment turned out. The conclusion might be that the hypothesis is correct. Sometimes, though, the hypothesis is not correct. In that case, the scientist might develop a new hypothesis and another experiment.

In the following experiments, we'll see the scientific method in action. First, we'll gather some facts and observations about light. For each experiment, we'll develop a question and a hypothesis. Next, we'll do an experiment to see if our hypothesis is correct. By the end of the experiment, we should know something new about light. Scientists, are you ready? Then let's get started!

Let's start observing light.

Experiment #1 Straight from the Sun

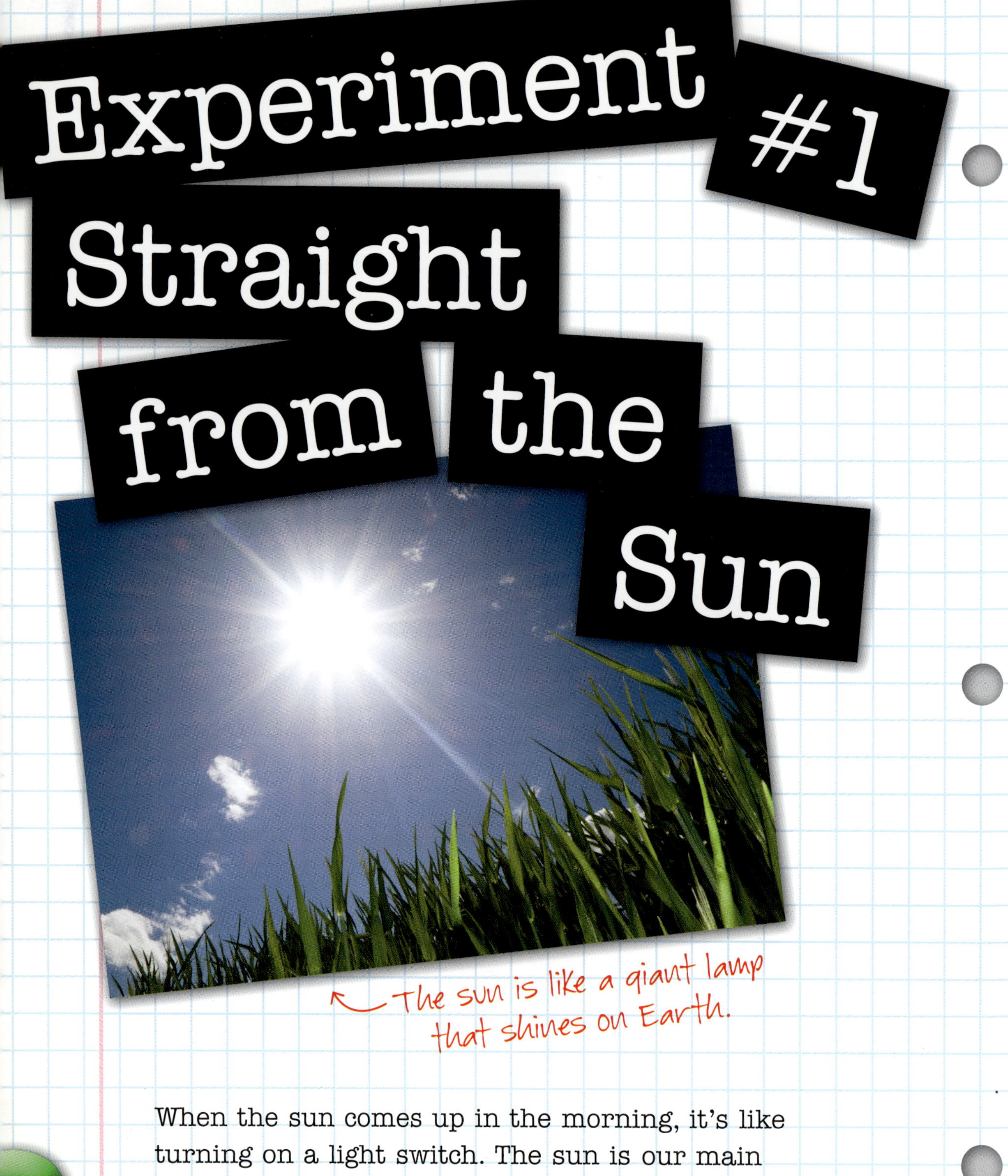

The sun is like a giant lamp that shines on Earth.

When the sun comes up in the morning, it's like turning on a light switch. The sun is our main source of light and helps us see the world.

Think of some other objects you have observed that give off light. A lamp, a flashlight, and even a birthday candle are all light sources.

Let's make some observations. When you're outside on a sunny day, light seems to be all around you. Inside, the light from a lamp seems to surround you, too. But how does light travel? We know it doesn't travel by bus or airplane! Perhaps a better question is: What type of path does light take?

Since light seems to surround us, our hypothesis could be: **Light does not travel in straight lines.** Let's see if this is correct.

Here are three sources of light. Can you name others?

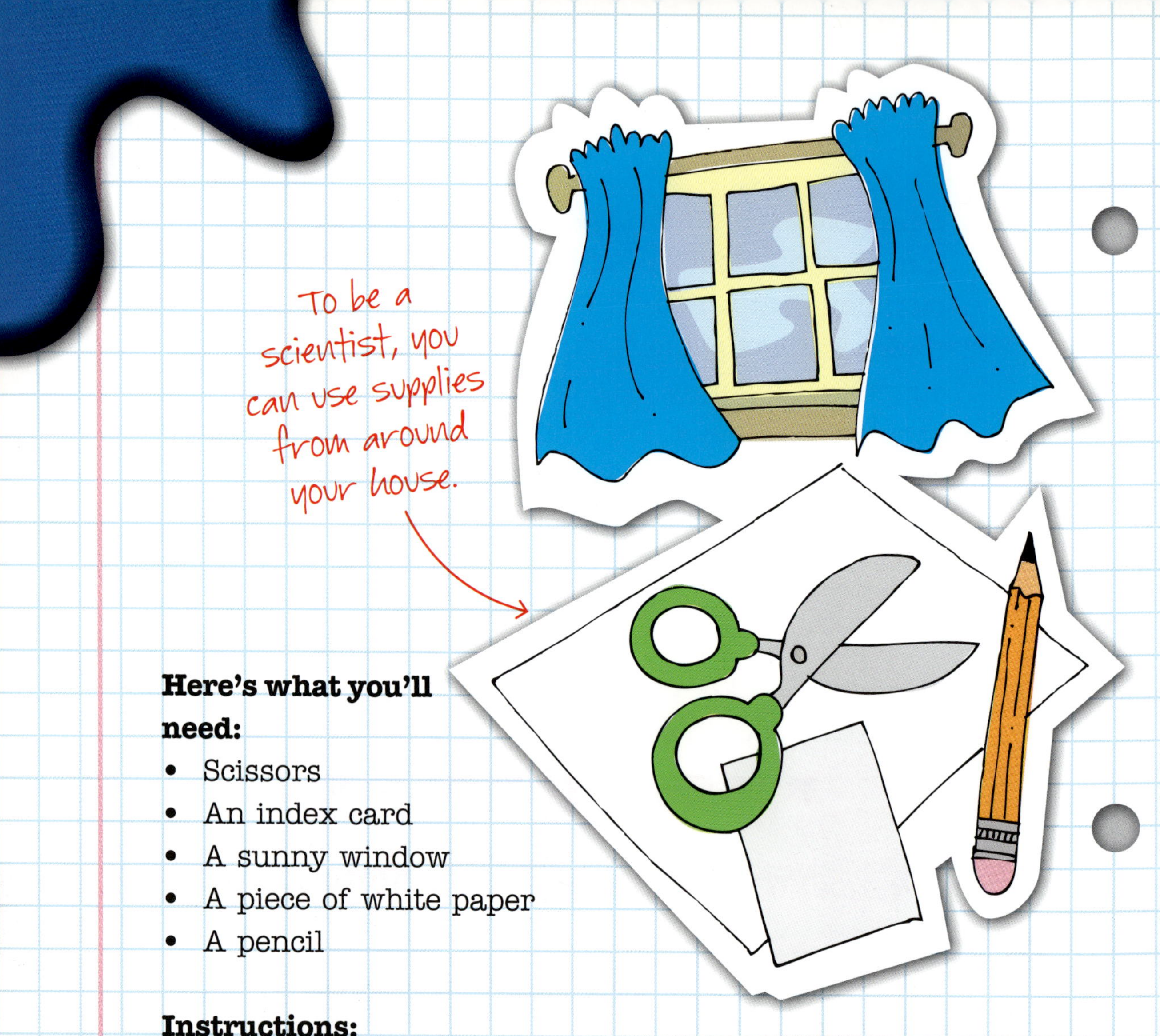

Here's what you'll need:

- Scissors
- An index card
- A sunny window
- A piece of white paper
- A pencil

Instructions:

1. Using scissors, cut out 10 slits along the side of an index card, so that it looks like a comb.
2. Stand near a sunny window.
3. Hold the card standing on its edge on top of a piece of white paper. Make sure the slits are pointing down. The card should face the sun so that the sunlight shines through the slits.
4. What do you see on the paper? Draw lines with a pencil where you see lines of light.

Conclusion:

What can you conclude from this experiment? Was our hypothesis correct? If light just surrounds us in space, traveling and bending in all directions, would you have expected to see straight lines on the paper? If light just surrounds us, it should have been able to go around the "teeth" of your paper comb and light up the whole paper. Instead, the light came straight through the slits.

Our hypothesis was not correct. Light does not just surround us. It travels in straight paths called rays. The lines from the sun are parallel. Parallel lines are straight and never touch. Sunlight always hits Earth in parallel paths. So why does it look like sunlight surrounds us? Because there are many straight lines of light traveling many straight paths. Rays of light from other sources, such as a flashlight or lamp, are not always parallel. But they are always straight.

Experiment #2 Shadows!

Your shadow follows you everywhere.

Walking down a sidewalk on a sunny day, you might see your shadow. It's shaped like you and either "walks" in front of you or follows behind. But what is a shadow? A shadow is created when a light source is blocked by an object. That object is often solid. It is also possible for gases and liquids to make shadows under the right conditions. Since light travels in straight lines, it can't bend around objects. But do all solid objects block light?

You can't see light through most objects. If you held a book over your face, the book would block all of the light and you wouldn't be able to see. But light does pass through some objects. If it didn't, people wouldn't be able to see through their eyeglasses! An object is called opaque when no light passes through it. An object is transparent when you can see light passing through it.

So you may wonder: Do transparent objects make shadows? Let's experiment. Here are two possible hypotheses:

Hypothesis #1: Transparent objects make shadows.

Hypothesis #2: Transparent objects do not make shadows.

Here's what you'll need:

- A teddy bear
- A table
- A desk lamp
- A clear, empty drinking glass

A teddy bear is opaque. A clear drinking glass is transparent.

Instructions:

1. Place the teddy bear on the table. Turn on the desk lamp and point it toward the teddy bear. Observe the shadow made by the teddy bear.
2. Move the lamp closer to the bear. Did the teddy bear's shadow change?
3. Place the drinking glass on the table. Turn the desk lamp on and point the light toward the glass. Observe whether or not the glass makes a shadow.
4. Move the lamp closer to the glass. Observe any changes. Write down your observations.

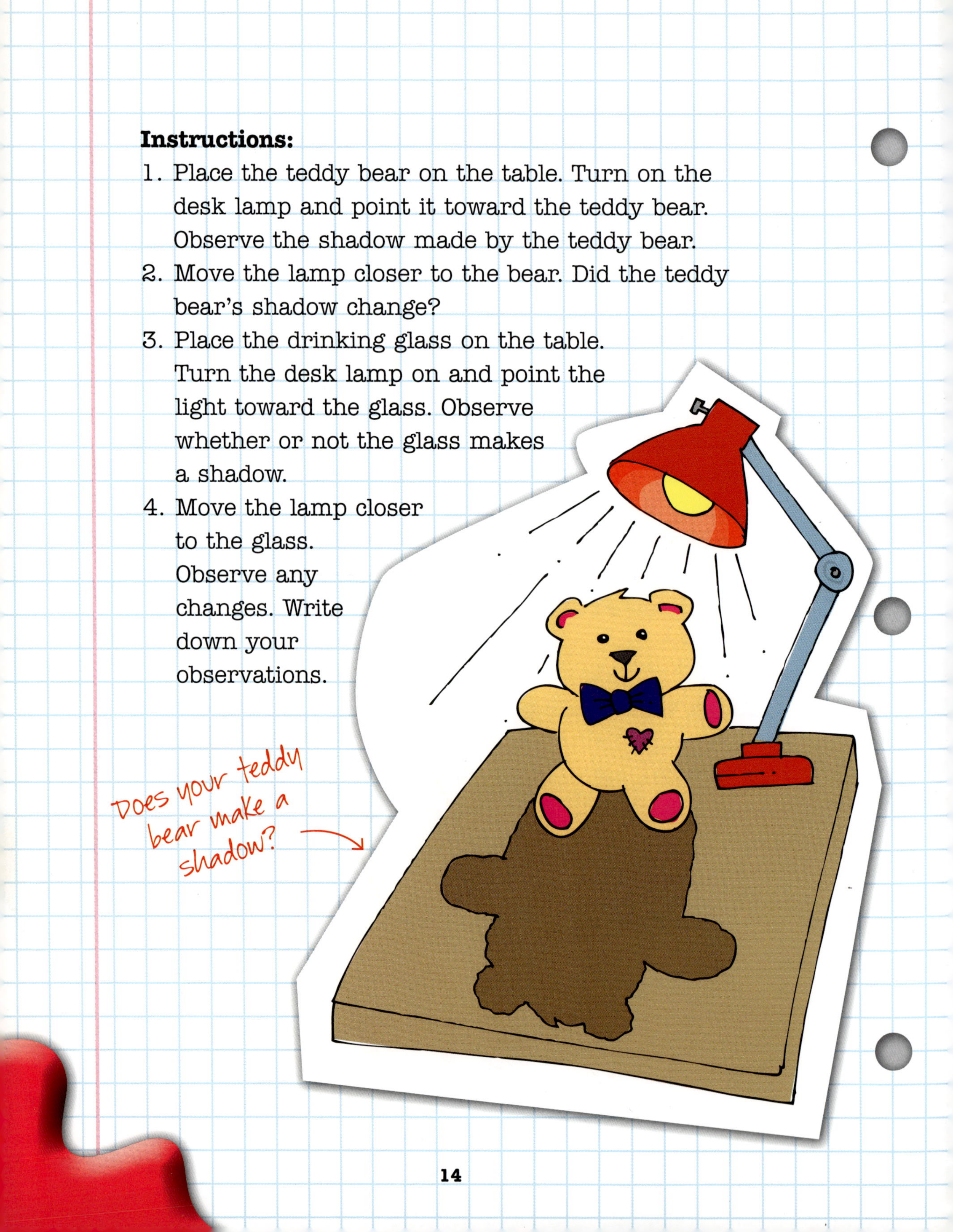

Conclusion:

The teddy bear is an opaque object. The drinking glass is transparent. What happens when you move the lamp shining on both of these objects?

What did you conclude from this experiment? Did both of the objects create shadows? Which hypothesis was correct? If the glass did create a shadow, was it the same type of shadow that the teddy bear made? How were the shadows different?

The drinking glass may not have made as sharp a shadow as the teddy bear, because light could pass through the glass. Even if it didn't block all of the light, would you say the glass blocked or changed some of the light? We'll see how glass changes light in another experiment.

You may notice that the middle part of a shadow is very dark. This part of the shadow is called the umbra. All light is completely blocked in this area. The fuzzier outline around the umbra is called the penumbra. It's the area that still lets some light around the edges of your object.

Experiment #3 Bouncing off the Wall

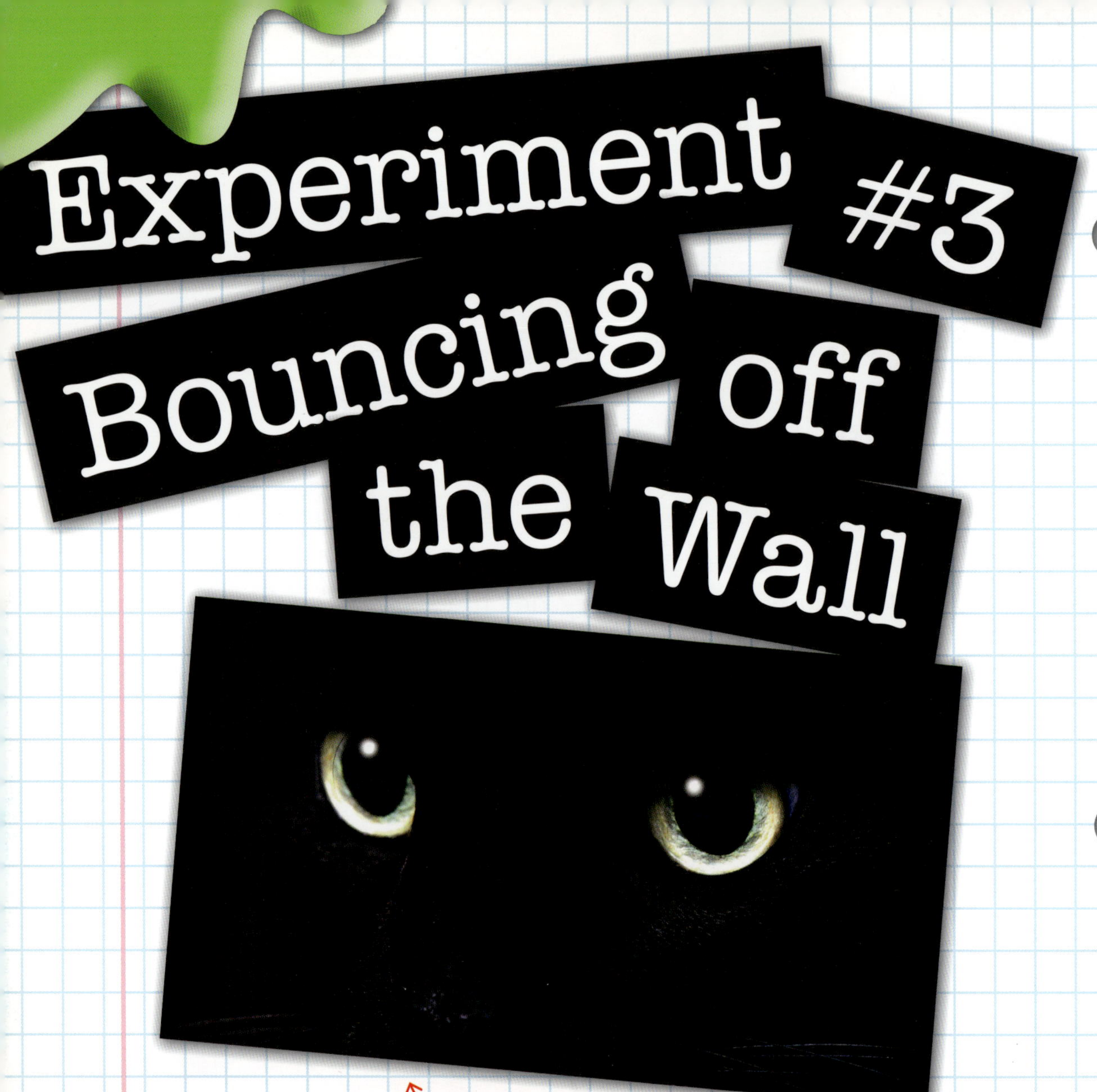

Cats can see better in the dark than people can.

From our first experiment, we know that light travels in straight lines. But can you change the path that light travels? Think about running. You can run in a straight line. But what happens when something blocks your path? If you run into that object, you'll bounce off of it. Did you know that light also bounces?

Bouncing light lets us see objects and colors. Light hits objects and bounces off them. We are able to see those objects when that reflected light reaches our eyes. Your eyes have trouble seeing in the dark. Eyes need light to work.

Look at the moon. Did you know that the moon doesn't give off its own light? Then why does it seem to shine? It actually reflects the light from the sun. The light comes from the sun and bounces off the moon down to Earth.

The moon does not give off any light of its own.

Mirrors also reflect light. When you look in a mirror, you see a reflection. That's because light bounces off your face and body. The light travels in a direct line to the mirror. Then it bounces off the mirror along the same line back toward your eyes.

But have you ever noticed that you don't need to be directly in front of a mirror to see a reflection? If you hide behind a doorway, you can hold out a mirror and see the room reflected in it, like a spy! How can we spy into a room that we are not in ourselves? We can do this because of the way light rays reflect off of mirrors. Let's test a new hypothesis: **If light hits a mirror at an angle, it will bounce off at an angle, too.**

Here's what you'll need:

- A mirror
- Masking tape
- A friend
- A flashlight

Instructions:

1. Set up a mirror on one side of a dark room. Stand on the other side of the room directly in front of the mirror. Mark this point on the floor with an X of masking tape.
2. Have a friend stand next to you. Turn on your flashlight, and shine it straight at the mirror. Can you spot which direction the reflected beam is heading?
3. Next, you and your friend should take two steps away from each other, in opposite directions from the X. Shine the flashlight at the mirror. Where is the beam of light now? Adjust your angle and try to get the beam to fall on your friend's shirt.
4. Now, both of you take another step away from each other. Can you keep the beam on your friend's shirt?

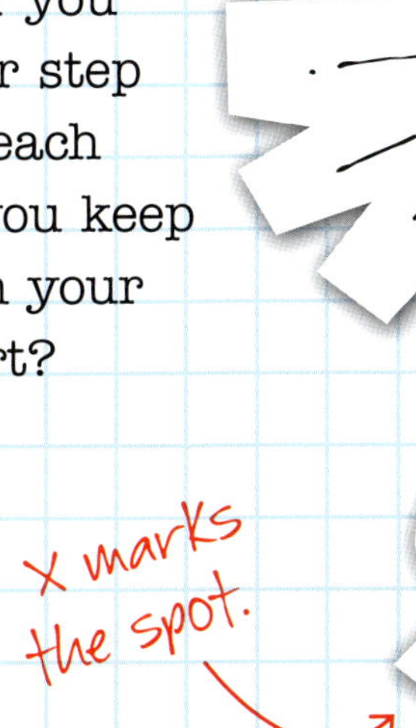

Conclusion:

How were you able to shine a light on the person next to you when you weren't pointing your flashlight at this person? What was the mirror doing to the light rays? How did the beam of reflected light change position as you moved away from the X? Why do you think it changed? What can you conclude about the way the light bounced off the mirror?

What happens to the reflected light when you move the flashlight?

As the angle of your light beam grew bigger, the angle it bounced off the mirror got bigger, too. The line that the light traveled to the mirror is called the **incident ray**. The path it took away from the mirror is the **reflected ray**. The angles of the incident ray and the reflected ray are always equal. That's called the law of reflection.

Experiment #4
Bent out of Shape

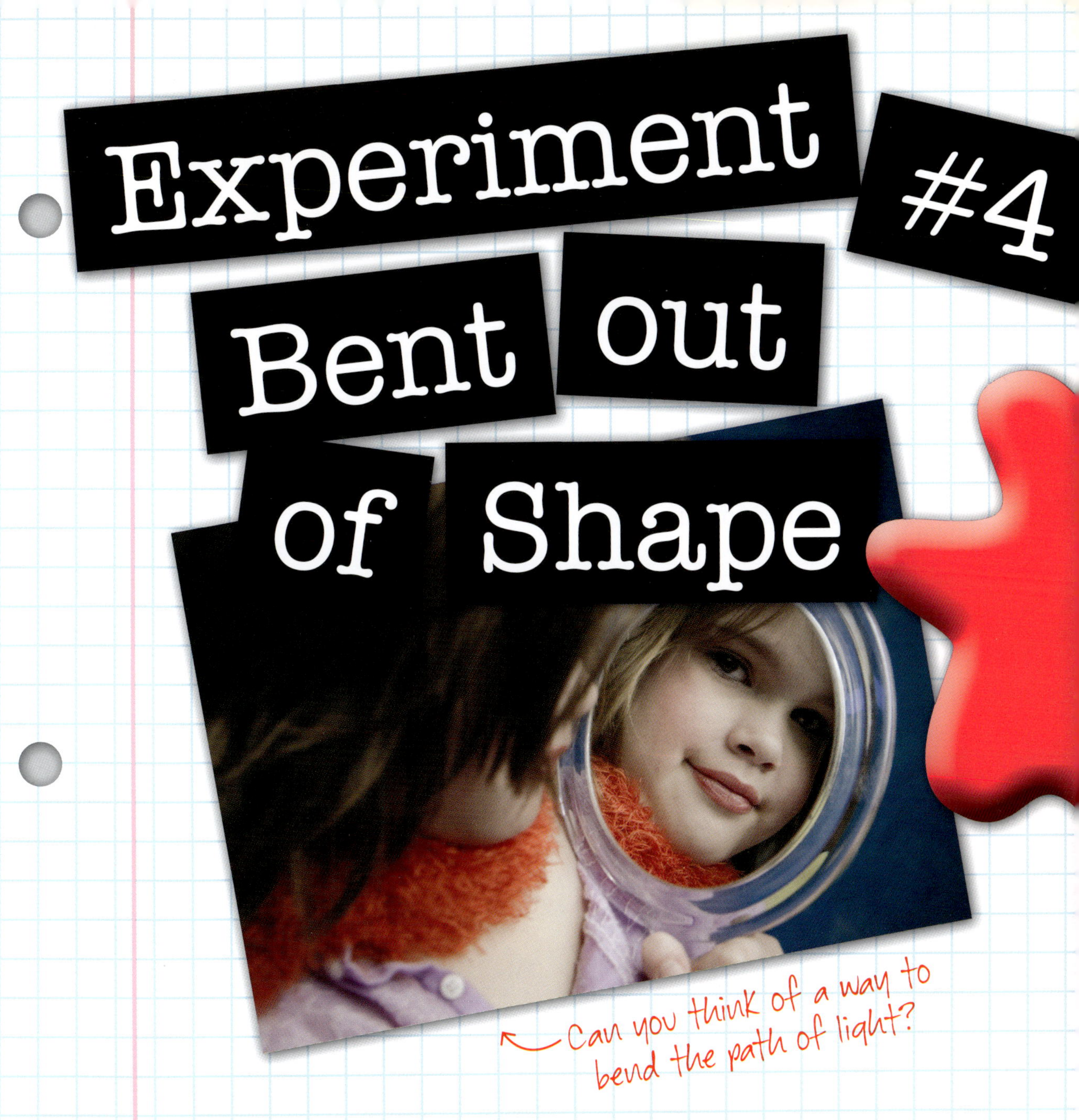

Can you think of a way to bend the path of light?

Light still traveled in a straight path when we bounced it off a mirror's surface. We know from our experiments that light travels in straight paths. What if we don't let it? Do you think we can bend the path of light?

We know from our past experiments that light sometimes travels through transparent objects or substances. The glass in Experiment #2 created a little shadow but let some light through. Somehow, objects like a glass can affect the path of light. Let's test another hypothesis: **The path of light will change when it passes through certain substances.**

Here's what you'll need:

- A tall, clear drinking glass
- Water
- A straw (or anything long, slim, and straight, such as a pencil)

Does your straw look like this?

Instructions:

1. Fill a glass about three-quarters full with water.
2. Place the straw in the glass at an angle.
3. Look at the glass from the side. Describe how the straw appears above the water. Describe how it appears below the water's surface. Move the straw straight up and down, and then at an angle again. How does it change? Write down what you observe.

4. Now look at the glass from slightly above so you can see the straw above the water, down through the surface, and under the water. Describe what you see.

Conclusion:

What happened to our straight straw? Did it appear bent or straight when we viewed it through water? Now pull out the straw. Do you see any breaks or bends? What can you conclude about the way light travels through water? Was our hypothesis correct?

Light travels through space at 186,282 miles per second (299,792 kilometers per second). That's really fast. The straw seems to bend in the water because light rays travel slower through water than through air. When light travels from one transparent substance to another, the light will bend as the speed of light slows down. We call this bending of light refraction. Think of some other transparent substances.

If something is translucent, some light can pass through it. But the light is scattered in different directions. You cannot see clearly through translucent objects. Waxed paper is an example of a translucent material.

Experiment #5
All the Colors in One

Rainbows are an amazing sight!

What color is light? You probably said white or yellow. Light is actually made up of lots of colors. In fact, it's made up of all the colors of the rainbow, plus more colors you can't even see!

Maybe you've spotted a rainbow after a thunderstorm in the summer. Maybe you've seen one in the spray of a fountain. You saw these rainbows because the air was filled with sunlight and water.

How can we prove that light is made up of many colors? Sometimes scientists use a tool called a prism as a way to bend light into its colors. A prism is a pyramid-shaped block of transparent glass or plastic. Since we know water bends light, and mirrors reflect light, perhaps we could combine them in a way to make our own prism. Let's test a hypothesis: **We can use water and a mirror to make a prism and break light into its rainbow colors.**

Here's what you'll need:

- A shallow baking pan
- Water
- A sunny window
- A mirror

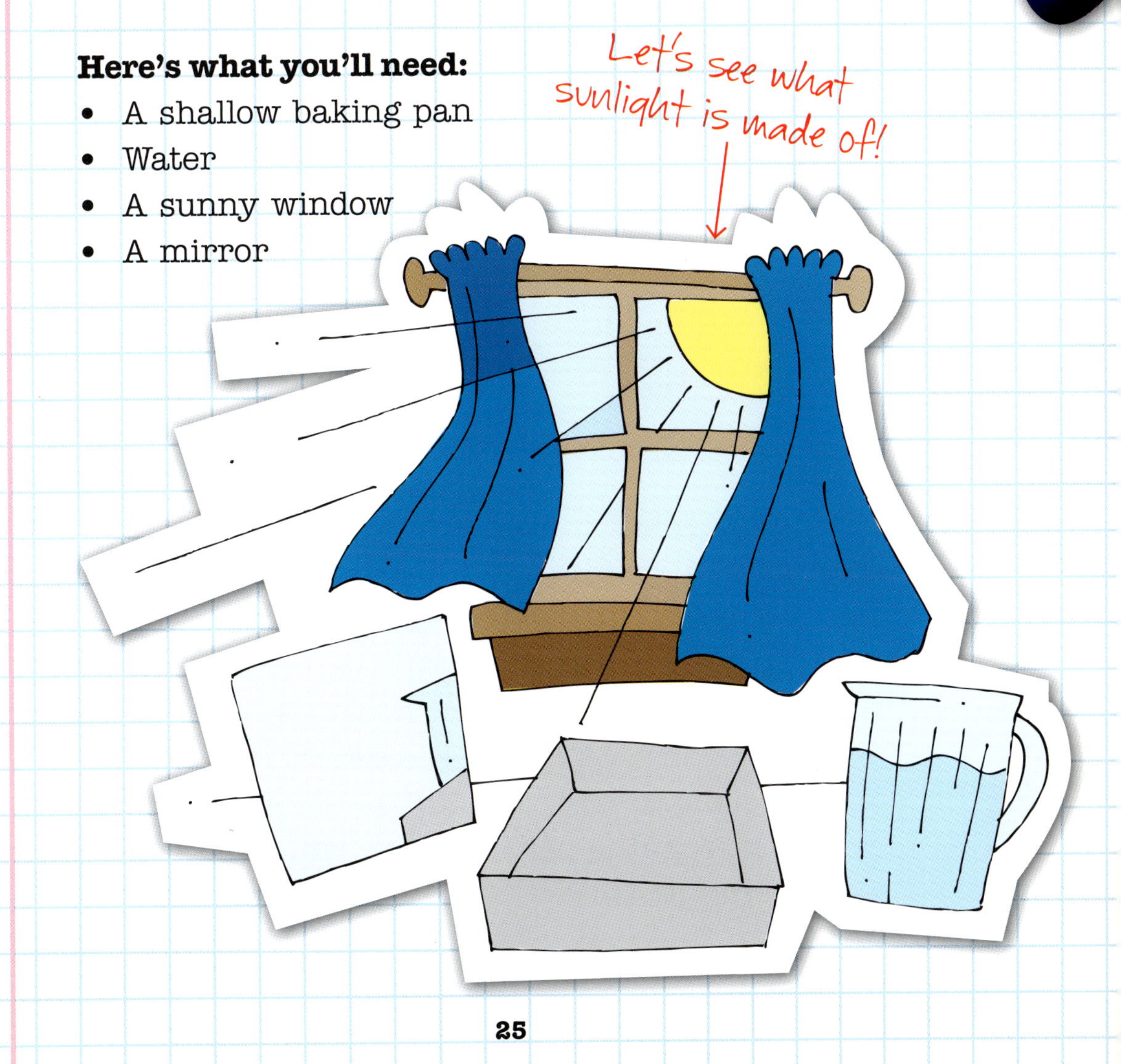

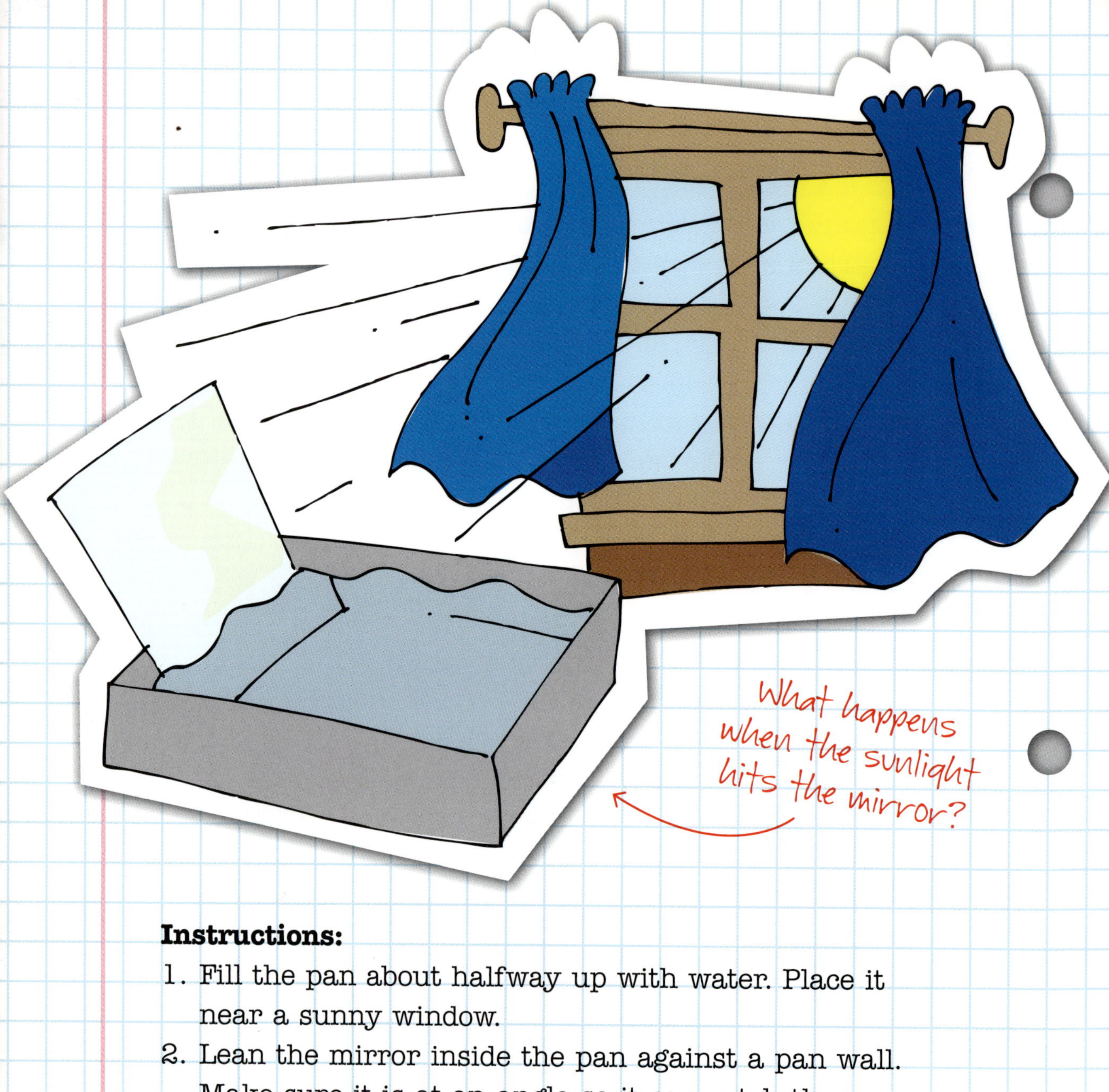

Instructions:

1. Fill the pan about halfway up with water. Place it near a sunny window.
2. Lean the mirror inside the pan against a pan wall. Make sure it is at an angle so it can catch the sunlight. Part of the mirror should be under water.
3. Adjust the mirror to make a reflection appear on the ceiling or on the wall. You want to see a reflection from the part of the mirror that is under water. What do you see on the ceiling or wall?

Conclusion:

How did you change the sunlight into a rainbow? Was our hypothesis correct? Where did the red, orange, yellow, green, blue, indigo, and violet colors come from? Did you know that the colors always fall in that same order?

We learned earlier that light rays travel in straight paths. Inside these paths, the light rays travel in a wave pattern. The distance between the tops of each wave is called a wavelength. Each color of the rainbow has a different wavelength. Violet has the shortest wavelength, and red has the longest wavelength.

Your water prism in this experiment refracted and reflected the light. Because violet has the shortest wavelength, it bent the most. Red bent the least. So parts of the sunlight came off the prism at different angles.

Light is a type of electromagnetic energy from the sun. This energy travels through space in different forms, depending on its wavelength. We can only see a few of these electromagnetic waves coming from the sun. The part we can see is called visible light. This light is made up of the colors of the rainbow. These rainbow colors are called the spectrum.

Experiment #6 Do It Yourself!

Hmmm . . . what other experiments can I do with light?

You can probably think of many more questions about light. When you brush your teeth, look in the mirror. You might be holding your toothbrush in your right hand. But why is the person in the mirror holding it with the left? Look at the back of a spoon. As you move it closer and farther away from your face, how does your reflection change? What makes it change that way? Why does your shirt look blue in the inside light, but more purple when you walk outside? Why do you seem to have no shadow when the sun is overhead?

Pick one of these questions, or come up with one of your own. Then find the answer by setting up your own experiment. What is your hypothesis? What items do you need to do your experiment? Write down how you will run your experiment. Do the experiment. What can you conclude from it?

Scientists always ask questions. It's easy to see why. The world around us is full of mysteries waiting to be solved. Choose one and create an experiment. You might come up with a bright new idea!

Do you have a spoon and some creative thoughts? Way to go, scientist!

GLOSSARY

conclusion (kuhn-KLOO-zhuhn) a final decision, thought, or opinion

electromagnetic energy (eh-LEK-troh-mag-NET-ik EN-uhr-jee) energy from the sun in the form of waves

hypothesis (hy-POTH-uh-sihss) a logical guess about what will happen in an experiment

incident ray (IN-sih-duhnt RAY) the path of light toward a surface

method (METH-uhd) a way of doing something

observations (ob-zur-VAY-shuhnz) things that are seen or noticed with one's senses

opaque (oh-PAYK) not able to let light pass through

reflected ray (ri-FLEK-tuhd RAY) the path of light that bounces off a surface

refraction (ri-FRAK-shuhn) the bending of light

translucent (trans-LOO-suhnt) able to let some light pass through, but scattering it in different directions

transparent (trans-PEHR-uhnt) able to let light pass through

FOR MORE INFORMATION

BOOKS

Clark, John Owen Edward. *Light and Sound.* Milwaukee, WI: Gareth Stevens Publishing, 2006.

Gardner, Robert. *Dazzling Science Projects with Light and Color.* Berkeley Heights, NJ: Enslow Elementary, 2006.

Richards, Jon. *Light & Sight.* New York: PowerKids Press, 2008.

WEB SITES

Exploratorium Science Snacks
www.exploratorium.edu/snacks/
Pictures and descriptions of neat exhibits related to light

Hunkin's Experiments: Experiments with Light
http://www.hunkinsexperiments.com/themes/themes_light.htm
Cartoons that show experiments you can do with light

Kids Science Experiments
www.kids-science-experiments.com/index.html
Fun experiments and activities, including several on bending light, reflecting light, and colors

INDEX

bending light, 11, 12, 21–23, 25
bending light experiment, 22–23
bouncing light experiment, 18–20

conclusions, 7, 11, 15, 20, 23, 27, 29

do-it-yourself experiment, 29

electromagnetic energy, 27
eyes, 16, 18

hypotheses, 6–7, 9, 11, 13, 15, 18, 22, 23, 25, 27, 29

incident rays, 20

law of reflection, 20

moon, 17

observations, 6, 7, 9, 14

opaque objects, 13, 15

parallel rays, 11
paths of light, 5, 9, 11, 17, 20, 21, 22, 27
penumbra, 15
prism experiment, 25–27

rays, 11, 18, 20, 23, 27
reflected light, 5, 16–20, 25, 26, 27, 28–29
reflected light experiment, 18–20
reflected ray, 20
refracted light, 5, 23, 27
refracted light experiment, 22–23

scientific method, 6–7
scientists, 4, 5, 6, 7, 25, 29
shadow experiment, 13–15
shadows, 12–15, 29
sources of light, 5, 8, 9, 11, 12
spectrum, 27
speed of light, 23
sun, 8, 9, 10, 11, 12, 17, 24, 26, 27, 29

translucent objects, 23
transparent objects, 13, 15, 22, 23, 25
travel (movement), 5, 9, 11, 12, 17, 18, 20, 21–23, 27
travel experiments, 9–11, 22–23

umbra, 15

wavelengths, 27

About the Author

Dana Meachen Rau loves to observe. Wondering about nature inspires her to write. Researching information helps answer her questions about how the world works. And writing books allows her to share what she has discovered with others. Mrs. Rau has written more than 200 books, many on science topics, for children of all ages. She lives, writes, and experiments at home in Burlington, Connecticut.